RIBA

Concise Conditions of Appointment for an

Architect

2010

2012 edition

Incorporating Amendment 1, September 2011 and Amendment 2, January 2012

This and the following eight pages (numbered 2–10) are the Concise Conditions of Appointment for an Architect 2010, 2012 edition, referred to in the Agreement relating to

The Project, namely:

between

The Client, namely:

Initials

and

The Architect, namely:

Initials

Royal Institute of British Architects

RIBA Agreements are produced in association with:

ACA	RIAS	RSAW	RSUA
Association of Consultant Architects	**Royal Incorporation of Architects in Scotland**	**Royal Society of Architects in Wales**	**Royal Society of Ulster Architects**

The components of a Concise Agreement 2010 are the Conditions of Appointment, schedules of Services and Fees and Expenses, any appendices and a Letter of Appointment.

A list of the principal changes from the 2010 Conditions can be found on the reverse of the pack cover sheet and is also available online at www.ribabookshops.com/agreements

The Concise Conditions of Appointment are applicable for a Client who is acting for business or commercial purposes or is a Public Authority. The Concise Conditions may also be used for a Client who is a consumer ie 'a natural person acting for purposes outside his trade, business or profession but the agreement will be subject to the Unfair Terms in Consumer Contracts Regulations 1999, which require the terms of the Agreement, these Conditions, the schedule of services, any appendices and the Letter of Appointment to be *individually negotiated*. See Concise Agreement 2010: Notes.

Individual architects are required to be registered with the Architects Registration Board, are subject to its Code and to the disciplinary sanction of the Board in relation to complaints of unacceptable professional conduct or serious professional incompetence.

All parties must rely exclusively upon their own skill and judgement or upon their advisers when using components from the suite of RIBA Agreements 2010, and neither RIBA nor RIBA Enterprises Ltd assumes any liability to any user or any third party.

- ✔ This document is for the sole use of the purchaser.
- ✔ You may use it for one professional services contract only.
- ✔ You and the other party to your contract may distribute copies of this publication to advisers and to other persons as necessary in connection with the proper performance of your contract.
- ✘ In any other case, you may not distribute or reproduce the whole or any part of this document in any work, whether in hard copy, electronic or any other form, without the prior written consent of RIBA Enterprises Ltd.

Notes on using components in PDF format (PDFs only)

Conditions of Appointment and guides are available as locked PDFs. They are copyright protected as literary works and cannot be edited. Amendments to the Conditions, if needed, are made by hand on the face of the document or in a separate appendix.

When downloaded from the website – www.ribabookshops.com/agreements Conditions of Appointment will include the project details – name of project, client and architect or consultant – in the identification box and as a footer on each page inserted by the purchaser online.

An agreement in electronic format will comprise the Conditions together with the core and/or other components that are also available online in Rich Text Format (RTF) as required. These components, eg schedules and notes and model letters, can be customised using most commonly used word-processing software, such as MS Word, to meet project requirements or modified to match the house style of the practice.

Copyright notice

© Royal Institute of British Architects, 2012
Concise Conditions of Appointment for an Architect 2010, 2012 edition

Published by RIBA Publishing, 15 Bonhill Street, London EC2P 2EA
First published 2010 updated September 2011 and January 2012

Printed and bound in Great Britain

RIBA Publishing is part of RIBA Enterprises Ltd.
www.ribaenterprises.com

Concise Conditions of Appointment for an Architect 2010, 2012 Revision

1 Definitions and general interpretation

1.1 **Collaborate** means to co-operate with and to provide to or receive from Other Persons, as and when requested, information reasonably necessary for performing and, where the Architect considers itself competent to do so, to comment on such information.

1.2 **Construction Act** means the Housing Grants Construction & Regeneration Act 1996 and the Local Democracy Economic Development and Construction Act 2009.

1.3 **Letter of Appointment** means the letter of appointment to which these Conditions are annexed.

Other Persons means any person, company or firm, other than the Architect or any sub-consultant of the Architect, including but not limited to consultants, contractors, sub-contractors, specialists, site inspectors or clerks of works, statutory bodies or undertakers, approving or adopting authorities, who have performed or will perform work or services in connection with the Project.

Project is defined in the Letter of Appointment.

Services means the services to be performed by the Architect specified in the Schedule of Services, which may be varied by agreement.

1.4 Where under this Agreement an action is required within a specified period of days from a specified date, that period commences immediately after that date. The period includes Saturdays and Sundays but excludes any day that is a public holiday.

1.5 The provisions of this Agreement are without prejudice to the respective rights and obligations of the parties and continue in force as long as necessary to give effect to such rights and obligations.

1.6 This Agreement is subject to the law and the parties submit to the exclusive jurisdiction of the courts of England and Wales or Northern Ireland or Scotland as specified in the Letter of Appointment.

2 Architect's Services

2.1 The Architect shall exercise reasonable skill, care and diligence in accordance with the normal standards of the Architect's profession in performing the Services and discharging all the obligations under this condition 2.

2.2 The Architect shall:

   2.2.1 perform the Services with due regard to the Client's requirements;

   2.2.2 advise on progress in the performance of the Services, of any information, decision or action required or of any issue that may materially affect the delivery, the cost or quality of the project;

   2.2.3 (a) act on behalf of the Client's in the matters set out or implied in this Agreement;
   (b) if acting as contract administrator of a building contract exercise impartial and independent judgement when dealing between the Client and the contractor;

   2.2.4 collaborate with any Other Persons who can reasonably be expected to be appointed by the Client and, as applicable, shall integrate relevant information received from such persons into the Architect's work;

   2.2.5 make no material alteration to the Services or the approved design without the consent of the Client, except in an emergency.

2.3 The Architect shall have the right to publish photographs of the Project, and the Client shall give reasonable access to the Project for this purpose for two years after practical completion of the construction works.

Concise Conditions of Appointment for an Architect 2010, 2012 Revision

3 Client's responsibilities

3.1 The Client:

    3.1.1 shall advise the Architect of the requirements and of any subsequent changes required;

    3.1.2 shall provide, free of charge, the information in the client's possession, or which is reasonably obtainable, and which is necessary for the proper and timely performance of the Services and the Architect shall be entitled to rely on such information;

    3.1.3 shall give decisions and approvals necessary for the performance of the Services;

    3.1.4 may issue reasonable instructions to the Architect.

3.2 The Client shall:

    3.2.1 appoint or otherwise engage any Other Persons required to perform work or services under separate agreements and shall require them to collaborate with the Architect;

    3.2.2 hold the contractor or contractors and not the Architect responsible for the proper carrying out and completion of construction works;

    3.2.3 not deal with the contractor or contractors directly or interfere with the Architect's duties or actions under the building contract.

3.3 The Client acknowledges that the Architect does not warrant:

    3.3.1 that planning permission and other approvals from third parties will be granted at all or, if granted, will be granted in accordance with any anticipated time-scale;

    3.3.2 compliance with any programme and/or any target cost for building work, which may need to be reviewed for, but not limited to:
(a) variations requested by the Client;
(b) variation in market prices;
(c) delays caused by any Other persons or any other factor beyond the control of the Architect;
(c) the discovery at any time of previously unknown conditions.

    3.3.3 the competence, performance, work, services, products or solvency of any Other Persons.

4 Assignment[1]

4.1 Neither the Architect nor the Client shall at any time assign the benefit of this Agreement or any rights arising under it without the prior written consent of the other, which consent shall not be unreasonably withheld or delayed.

4.2 The Architect shall not sub-contract performance of any part of the Services without the prior consent of the Client, which consent shall not be unreasonably withheld or delayed.

5 Fees and expenses

5.1 The fees for performance of the Services and/or any additional services shall be calculated in accordance with this condition 5 and as specified in the Letter of Appointment.

5.2 The Basic Fee for performance of the Services shall be:

    5.2.1 the specified percentage applied to the actual cost of the building work; or

---

[1] Assignation in Scotland.

- 5.2.2 the separate percentages specified for each work stage applied to the approved cost of the building work at the end of the previous stage; or

- 5.2.3 the specified fixed lump sum or sums; or

- 5.2.4 time charges ascertained by multiplying the time reasonably spent in the performance of the Services by the specified hourly or daily rate for the relevant personnel. Time 'reasonably spent' includes the time spent in connection with performance of the Services in travelling from and returning to the Architect's office; or

- 5.2.5 any combination of these; and/or

- 5.2.6 any other agreed method.

Where a percentage fee applies, the percentages are applied to the current approved estimate of cost of the building works or the contract sum. The cost shall exclude VAT, fees and any claims made by or against the contractor or contractors.

5.3 Lump sums, rates for time charges, mileage and printing shall be revised every 12 months in accordance with changes in the Consumer Price Index. Each 12-month period commences on the anniversary of the date on which the Architect commenced performance of the Services.

5.4 The Basic Fee shall be adjusted:

- 5.4.1 including due allowance for any loss and/or expense, if material changes are made to the Brief and/or the latest approved estimate of the cost of the building work and/or the programme save to the extent that any changes arise from a breach of this Agreement by the Architect; and/or the Services are varied by agreement;

- 5.4.2 where percentage fees in accordance with conditions 5.2.1 or 5.2.2 apply, to compensate for any reduction of the Construction Cost arising solely from deflationary market conditions not prevailing at the date of this Agreement.

5.5 If the Architect is involved in extra work or incurs extra expense for reasons beyond the Architect's reasonable control, additional fees shall be calculated on a time basis in accordance with condition 5.2.4 where:

- 5.5.1 the cost of any work, installation or equipment, for which the Architect performs Services, is not included in the cost for the building work; and/or

- 5.5.2 the Architect is required to vary any item of work commenced or completed or to provide a new design after the Client has authorised development of an approved design; and/or

- 5.5.3 performance of the Services is delayed, disrupted or prolonged.

The Architect shall inform the Client on becoming aware that this condition 5.5 will apply. This condition 5.5 shall not apply to the extent that any change or extra work or expense arises from a breach of this Agreement by the Architect.

5.6 The Client shall reimburse the Architect for expenses in the manner specified in the Letter of Appointment.

5.7 The Architect shall maintain records of time spent on Services performed on a time basis and for any expenses and disbursements to be reimbursed at net cost. The Architect shall make such records available to the Client on reasonable request.

Concise Conditions of Appointment for an Architect 2010, 2012 Revision

Payment notices

5.8 [2] The Architect shall issue payment notices at the intervals specified in the Letter of Appointment.

Each notice shall comprise the Architect's account setting out any accrued instalments of the fee and other amounts due, less any amounts previously paid and stating the basis of calculation of the amount specified as which shall be "the notified sum". The payment due date shall be the date of the Architect's Payment Notice. Instalments of fees shall be calculated on the Architect's reasonable estimate of the percentage of completion of the Services or stages or other services or any other specified method.

The Client shall pay the notified sum within 14 days of the date of issue of the relevant Notice (which shall be the "final date for payment") unless:

(a) The Architect has become insolvent (as defined in the Construction Acts at any time between the last date on which the Client could have issued the Notice under 5.9 and the final date for payment);

(b) The Client issues a notice under 5.9.

Otherwise the amount due and payable shall be the notified sum. The Client shall not delay payment of any undisputed part of the notified sum.

The Architect shall submit the final account for fees and any other amounts due when the Architect reasonably considers the Services have been completed.

Notice of Intention to Pay Less

5.9 If the Client intends to pay less than the notified sum the Client shall give a written notice to the Architect not later than 5 days before the final date for payment specifying the amount that the Client considers to be due on the date the Notice is served, the basis on which that sum is calculated and, if any sum is intended to be withheld, the ground for doing so or, if there is more than one ground, each ground and the amount attributable to it. The Client shall on or before the final date for payment make payment to the Architect of the amount if any specified in the written Notice.

If no such Notice is given the amount due and payable shall be the notified sum stated as due in the Architect's account. The Client shall not delay payment of any undisputed part of the account. If the Client issues such a Notice and the matter is referred to an Adjudicator who decides that an additional sum greater than the amount stated in the Notice of Intention to Pay Less is due, the Client shall pay that sum within 7 days of the date of the decision or the date which apart from the Notice would have been the final date for payment.

5.10 The Client shall not withhold any amount due to the Architect under this Agreement unless the amount has been agreed with the Architect or has been decided by any tribunal to which the matter is referred as not being due to the Architect.

All rights of set-off at common law or in equity which the Client would otherwise be entitled to exercise are expressly excluded.

5.11 If performance of any or all of the Services and/or obligations is suspended or ended, the Architect shall be entitled:

5.11.1 to payment of any part of the fee and other amounts properly due to the date of the last instalment and a fair and reasonable amount up to the date of termination or suspension, payment of any licence fee due under clause 6; together with

5.11.2 reimbursement of any loss and/or damages caused to the Architect by reason of the suspension or the termination, except where the Architect is in material or persistent breach of the obligations under the Agreement.

---

[2] In the event of non-payment of any amount properly due to the Architect under this Agreement, the Architect is entitled to interest on the unpaid amounts under the provisions of condition 5.12, may suspend use of the licence under the provisions of condition 6, may suspend or terminate performance of the Services and other obligations under the provisions of clause 8, or may commence dispute resolution procedures and/or debt recovery procedures.

RIBA

Small Project Services

Schedule 2010 (2012 revision)

This is the Small Project Services Schedule 2010 (2012 revision) referred to in the Agreement relating to

The Project, namely:

between

The Client, namely:

Initials

and

*

Initials

---

* Insert 'The Architect' or 'The Consultant', namely

Identification Box
The identification box above is not always necessary. For example, if you are incorporating this schedule into a document which already has an identification box, there is no need to include it.

However, if you decide to use it, you should amend the text so that it is clear what project it relates to.

Note that if you are adding this document as an Appendix to your Agreement, you should use the identification box to say so. For example, you could replace the top line of text with 'This is Appendix <insert reference> referred to in the Agreement relating to:'.

Alternatively, where the law of Scotland applies, amend the top line to read 'This and the following <number> pages (numbered 2 to <number>) is the <accurate title of document> referred to in the Agreement relating to:'

If used with the Sub-consultant Agreement, 2012 revision, amend 'The Client' to 'The Consultant' and 'The Architect' or 'The Consultant' to 'The Sub-consultant'.

Royal Institute of British Architects

Small Project Services Schedule, 2012 revision

Delete or strike through any services not required. Check the boxes for the required services [Y or ✔], alternatively enter 'T' for time-based services or 'LS' for lump sums, make any necessary amendments and/or add additional services.

The Services shall be performed in accordance with this Schedule in stages:

BEFORE CONSTRUCTION

AB Preparation
- ☐ Visit the property and carry out an initial appraisal
- ☐ Assist the Client in preparing the Client's requirements
- ☐ Discuss alternative solutions for the project
- ☐ Advise on the need for services by consultants or specialists

Delete (a) or (b)
- ☐ (a) Make (b) Arrange: survey of site and/or buildings
- ☐ Arrange investigation of soil or structural conditions
- ☐

CD Design
- ☐ Prepare a preliminary design and discuss with the Client
- ☐ Develop the final design
- ☐ Provide information for an approximate estimate of cost
- ☐ Submit the final design proposals and approximate cost for approval
- ☐ Make an application for detailed planning permission
- ☐

EFG Construction Information
- ☐ Co-ordinate and integrate any designs provided by others
- ☐ Prepare drawings and other information in sufficient detail to enable a tender or tenders to be obtained
- ☐ Make an application for Building Regulations approval

Delete (a) or (b)
- ☐ Prepare (a) a specification (b) a schedule of works
- ☐ Advise on an appropriate form of building contract, its conditions and the responsibilities of the Client, the consultants and the builder
- ☐

Small Project Services Schedule, 2012 revision

CONSTRUCTION

H Tender Action

☐ Prepare documents required for tendering purposes

☐ Advise on potential builders to be invited to tender for the work

☐ Invite, appraise and report on tenders

☐

JK Construction Work

☐ Advise on the appointment of a builder

☐ Prepare the building contract and arrange for it to be signed

☐ Provide the builder with information required for construction

☐ Visit the site to see that the work is proceeding generally in accordance with contract

☐ Certify payments for work carried out or completed. Advise on actual cost

☐ Provide or obtain record drawings showing the building and its services

☐ Give general advice on maintenance

☐

L After Handover

☐ Make final inspections and advise on resolution of any defects

☐ Agree final account and issue a final certificate

☐

Other Services

☐

☐

☐

☐

The following activities do not form part of the Services unless identified as 'Other Services' above:

- Models and special drawings
- Negotiating approvals by statutory authorities
- Making submissions to and negotiating approvals by landlords, freeholders, etc
- Preparing a schedule of dilapidations
- Services in connection with party wall negotiations
- Negotiating a price with a builder (in lieu of tendering)
- Dealing with extensions of time and contractor's claims
- Services in any dispute between the Client and another party
- Services following damage by fire and other causes
- Services following suspension, termination of any contract or agreement with or the insolvency of any other party providing services to the project
- Services in connection with government and other grants

Small Project Services Schedule, 2012 revision

What is this document?
This schedule covers the design and construction stages and any Other Services. As it is not profession-specific, it may be used for designers in any profession. The associated management roles eg Lead Consultant, Lead Designer, Contract Administrator etc are not specified in detail. If the Architect/Consultant is to perform any of them, the Role Specifications part of Services 2010 will provide appropriate guidance.

In this updated edition, the changes are limited to these explanatory notes.

Copyright Licence:
- ✔ This document is for the sole use of the purchaser.
- ✔ You may edit and amend it to match the style and procedures of your practice and as required for specific projects.
- ✔ You and the other parties to your contracts may distribute copies of this document to advisers and to other persons as necessary in connection with the proper performance of your contracts.
- ✘ In any other case, you may not distribute or reproduce the whole or any part of this document in any work, whether in hard copy, electronic or any other form, without the prior written consent of RIBA Enterprises Ltd.
- ✘ You may not use the RIBA initials or logo unless you are an RIBA chartered member or RIBA chartered practice.

Further guidance
A *Guide to RIBA Agreements 2010 (2012 revision)* explains fully how the RIBA Agreements 2010, 2012 revision work and their legal context.

Copyright notice

© **Royal Institute of British Architects 2010, 2012**
Small Project Services Schedule, 2012 revision

Published by RIBA Publishing, 15 Bonhill Street, London EC2P 2EA
First published 2010; updated September 2011
Revised edition published 2012

RIBA Publishing is part of RIBA Enterprises Ltd.
www.ribaenterprises.com

Concise Conditions of Appointment for an Architect 2010, 2012 Revision

5.12 In the event that any amounts are not paid when properly due, the payee shall be entitled to simple interest on such amounts until the date that payment is received at 8% per year over the dealing rate of the Bank of England Rate current at the date that payment becomes overdue, together with such costs reasonably incurred by the payee (including costs of time spent by principals, employees and advisors) in obtaining payment of any sums due under this Agreement.

5.13 The Client or the Architect shall pay to the other party who successfully pursues, resists or defends any claim or part of a claim brought by the other:

  5.13.1 such costs reasonably incurred (including costs of time spent by principals, employees and advisors) where the matter is resolved by negotiation or mediation; or

  5.13.2 such costs as may be determined by any tribunal to which the matter is referred.

5.14 In addition to the fees and expenses, the Client shall pay any Value Added Tax chargeable on the Architect's fees and expenses.

6 Copyright licence

6.1 The Architect shall own all intellectual property rights including the copyright in the drawings and documents produced in performing the Services and generally asserts the Architect's moral rights to be identified as the author of such work.

No part of any design by the Architect may be registered[3] by the Client without the written consent of the Architect.

Providing that all fees and/or other amounts properly due are paid, the Client shall have a licence to copy and use the drawings and documents only for purposes related to construction of the Project or its subsequent use or sale but may not be used for reproduction of the design for any part of any extension of the Project or any other project.

Copying or use of the drawings and documents by an Other Person providing services to the Project shall be deemed to be permitted under a sub-licence granted by the Client, whether such Material was issued by the Client or on the Client's behalf.

The Architect shall not be liable for any use of the drawings and documents other than for the purpose for which they were prepared.

7 Liability and insurance

Architect's liability

7.1 No action or proceedings arising from the failure of the Architect to keep to this Agreement shall be commenced after the expiry of six years from the date of the last Services performed under this Agreement or, if earlier, practical completion of construction of the Project or such earlier date as prescribed by law.

7.2 In any such action or proceedings:

  7.2.1 the Architect's liability for loss or damage shall not exceed the amount of the Architect's professional indemnity insurance available specified in the Letter of Appointment, providing the Architect has notified the insurers of the relevant claim or claims as required by the terms of such insurance.

  7.2.2 No employee of the Architect or any agent of the Architect shall be personally liable to the Client for any negligence, default or any other liability whatsoever arising from performance of the Services.

---

[3] Under the Registered Designs Regulations 2001.

Concise Conditions of Appointment for an Architect 2010, 2012 Revision

7.3 Without prejudice to the provisions of condition 7.2.1, the liability of the Architect shall not exceed such sum as it is just and equitable for the Architect to pay having regard to the extent of the Architect's responsibility for the loss and/or damage in question and on the assumptions that:

    7.3.1 all other consultants, contractors and Other persons providing work or services for the Project have provided to the Client contractual undertakings on terms no less onerous than those of the Architect under this Agreement;

    7.3.2 there are no exclusions of or limitations of liability nor joint insurance or co-insurance provisions between the Client and any other person referred to in this condition; and

    7.3.3 all the persons referred to in this condition have paid to the Client such sums as it would be just and equitable for them to pay having regard to the extent of their responsibility for that loss and/or damage.

7.4 The Architect shall maintain until at least the expiry of the period specified in condition 7.1 professional indemnity insurance with a limit of indemnity not less than the amount or amounts specified in the Letter of Appointment, provided such insurance continues to be offered on commercially reasonable terms to the Architect at the time when the insurance is taken out or renewed.

The Architect, when reasonably requested by the Client, shall produce for inspection a broker's letter or certificate confirming that such insurance has been obtained and/or is being maintained.

7.5 Except for the rights conferred by condition 7.2, nothing in this Agreement confers or is intended to confer any right to enforce any of its terms on any person who is not a party to it, other than lawful assignees.

8 Suspension or termination

8.1 The Client may suspend or end performance of any or all of the Services and other obligations by giving at least 7 days' written notice and stating the reason for doing so.

The Architect may suspend or end performance of any or all of the Services and other obligations by giving at least 7 days' written notice and stating the ground or grounds on which it is intended to do so. Such ground or grounds include, but are not limited to the Client's failure to pay any fees or other amounts due by the final date for payment, unless, where applicable, the Client has given effective notice under condition 5.9 of the intention to pay less than the amount stated in an Architect's account.

If the reason for a notice of suspension arises from a default:

    8.1.1 which is remedied, the Architect shall resume performance of the Services and other obligations within a reasonable period; or

    8.1.2 which is not remedied by the defaulting party the Agreement will end by giving at least 7 days' further written notice.

Where Services are suspended by either party and not resumed within three months, the Architect has the right to treat performance of the Services affected as ended on giving at least seven days' further written notice to the Client.

Concise Conditions of Appointment for an Architect 2010, 2012 Revision

9 Dispute resolution[4]

9.1 In the event of any dispute or difference arising under the Agreement, the parties may attempt to settle the matter by negotiation or mediation or as specified in the Letter of Appointment.

Adjudication

9.2 Either party may give notice at any time of the intention to refer a dispute or difference to an adjudicator:

    9.2.1 Referral of the dispute to such adjudicator shall be made within seven days of such notice.

    9.2.2 The appointment of the adjudicator shall be made in accordance with the procedures identified in the Project Data.

    9.2.3 The parties may agree who shall act as adjudicator or, the adjudicator shall be a person nominated at the request of either party by the nominator specified in the Project Data.

    9.2.4 The adjudicator may allocate between the parties the costs relating to the adjudication, including the fees and expenses of the adjudicator, in accordance with the provisions of condition 5.13.

9.3 The provisions for arbitration are:

    9.3.1 Without prejudice to any right of adjudication, where in the Letter of Appointment an arbitration agreement is made and either party requires a dispute or difference (except in connection with the enforcement of any decision of an adjudicator) to be referred to arbitration then that party shall serve on the other party a notice of arbitration to that effect and the dispute or difference shall be referred to a person to be agreed between the parties or, failing agreement within 14 days of the date on which the notice is served, a person appointed by the appointor specified in the Letter of Appointment on the application of either party.

    9.3.2 Where the law of England and Wales or Northern Ireland is the applicable law:
    (a) the Client or the Architect may litigate any claim for a pecuniary remedy which does not exceed £5,000 or such other sum as is provided by order made under section 91 of the *Arbitration Act 1996*;
    (b) in such arbitration the Construction Industry Model Arbitration Rules (CIMAR) current at the date of the reference shall apply;
    (c) the arbitrator shall not have the power referred to in Section 38(3) of the Arbitration Act 1996.

    9.3.3 Where the law of Scotland is the applicable law such arbitration shall be conducted in accordance with the provisions of the Arbitration (Scotland) Act 2010. [5]

---

[4] The Architect is expected to operate in-house procedures to promptly handle complaints and disputes relating to specific project or performance matters.

[5] The parties will need to consider whether any of the default rules in the Arbitration (Scotland) Act are to be modified or identified as not applicable.

Concise Conditions of Appointment for an Architect 2010, 2012 Revision

10	**Consumer's right to cancel:**[6]

10.1 The consumer Client has the right to cancel this Agreement for any reason by delivering or sending (including by electronic mail) a cancellation notice to the Architect at any time within the period of fourteen days starting from the date when this Agreement was made.

10.2 The notice of cancellation is deemed to be served as soon as it is posted or sent to the Architect or in the case of an electronic communication on the day it is sent to the Architect.

10.3 If the Architect was instructed to perform any services before the Agreement was signed by the Client or before the end of the fourteen day period and the instruction or instructions were confirmed in writing, the Architect shall be entitled to any fees and expenses properly due before the Architect receives the notice of cancellation.

10.4 The notice is to be addressed to the Architect and state:

The Client <name> hereby gives notice that the RIBA Concise Agreement 2010 with the Architect <insert name> and signed [on our behalf] by <name of person(s) who [will sign] [signed]> on <date of signing> is cancelled.

Client signature(s):

<address>

<date>

---

[6] This condition applies where the Project relates to work to the Client's home or a second home including a new home and the Client is a consumer who is acting for purposes outside his trade, business or profession and has signed this Agreement in his/her own name, ie not as a limited company or other legal entity.

RIBA Bookshops

Dear Customer,

Updated schedules included in your pack

Thank you for purchasing an RIBA Agreement from RIBA Bookshops.

Following the introduction of the RIBA Plan of Work 2013, certain documents within the RIBA Agreements suite have been updated. Therefore, we have included the following new document in your package:

Small Project Services Schedule 2013
(RIBA Plan of Work 2013 compatible version)

This new document has been prepared to be compliant with the RIBA Plan of Work 2013, which was launched by the RIBA on 21st May 2013. It has been supplied in addition to the Small Project Services Schedule within the *Concise Agreement 2010 (2012 Revision): Conditions of Appointment for an Architect* or the *Domestic Agreement 2010 (2012 Revision): Conditions of Appointment for an Architect* in your pack, so that you have the option of using either document.

Yours faithfully,

Tim Scudder
Marketing Coordinator, RIBA Bookshops & Publishing

p.s. For more details of the RIBA Plan of Work 2013, please visit: ribaplanofwork.com

RIBA

Fees and Expenses Schedule 2010
(2012 revision)

This is the Fees and Expenses Schedule 2010 (2012 revision) referred to in the Agreement relating to

The Project, namely:

between

The Client, namely:

Initials

and

*

Initials

* Insert 'The Architect' or 'The Consultant', namely

Royal Institute of British Architects

What is this document?
This schedule is for detailing the fees and expenses for the project and may be used with any of the RIBA Agreements 2010, 2012 revision. Where used with the Standard Agreement, 2012 revision, it is an essential component if the Memorandum of Agreement is used to execute the contract. However, it is merely optional if used in conjunction with any of the Model Letters, which already allow you to detail the fees and expenses for the project.

All parties must rely exclusively upon their own skill and judgement or upon their advisers when using electronic components from the suite of RIBA Agreements 2010, 2012 revision, and RIBA and RIBA Enterprises Ltd do not assume any liability to any user or any third party.

Copyright Licence:
- ✔ This document is for the sole use of the purchaser.
- ✔ You may use it for one professional services contract only.
- ✔ You and the other party to your contract may distribute copies of this document to advisers and to other persons as necessary in connection with the proper performance of your contracts.
- ✘ In any other case, you may not distribute or reproduce the whole or any part of this document in any work, whether in hard copy, electronic or any other form, without the prior written consent of RIBA Enterprises Ltd.
- ✘ You may not use the RIBA initials or logo unless you are an RIBA chartered member or RIBA chartered practice.

Further guidance
In this updated edition the layout of tables has been amended.

Copyright notice

© **Royal Institute of British Architects, 2010, 2012**
Fees and Expenses Schedule, 2012 revision

Published by RIBA Publishing, 15 Bonhill Street, London EC2P 2EA
First published 2010; updated September 2011
Revised edition published 2012

RIBA Publishing is part of RIBA Enterprises Ltd.
www.ribaenterprises.com

Fees and Expenses Schedule, 2012 revision

Basic Fee *Identify the applicable clauses, relevant work stages and the basis for calculation of the amount of the fee*

Stage   Clause   Fee   Notes

In stages J, K and L1 the Basic Fee includes ☐ site visits[1]

Unless otherwise stated, the maximum number of visits will be one per month

Other fees *Other fees are fees for any Other or Special Services required but not included in the Basic Fee and, if applicable, any specified copyright licence fee for the Client's use after the date of the last Service performed under this Agreement of materials produced by the Architect (drawings, documents etc.) in performing the Services.*

Subject   Clause   Fee   Notes

---

[1] Unless time charges apply in which case, delete.

Fees and Expenses Schedule, 2012 revision

Time Charges

Person/grade	Rate [2]	Person/grade	Rate [2]

Expenses and Disbursements

The specified expenses listed below will be charged [3]

a) at net cost plus a handling charge of ☐ % of net cost

or

b) by the addition to the amount due of ☐ % of the total fee

or

c) Other [4] ☐

Specified Expenses

☐

Other Expenses

(including disbursements [5]) will be charged at net cost plus ☐ % of net cost

Where applicable, travel will be charged at ☐ per mile

Hard copies of Drawings and Documents

	A4	A3	A2	A1	A0
in black and white					
in PDF format or similar					

Payment

VAT Registration number of the payee is: ☐

Accounts for instalments of fees [6] shall be issued and paid:

☐ monthly ☐ at the end of each Work Stage

Delete or amend as required
Where payments are to be equalised or based on milestones, specify details:

---

[2] £/hour or £/day.
[3] Delete those that do not apply.
[4] Use where specified expenses are to be charged by a method different from those listed above.
[5] Such as payments to the Local Authority for planning submissions or Building Regulations approval.
[6] State whether calculated on basis of the estimate of the percentage of completion of the Services or stages or additional services or such other specified method.

Small Project Services Schedule, 2013

What is this document?
This schedule covers services in the Before Construction, Construction and After Construction stages, set out in accordance with the RIBA Stages in the RIBA Plan of Work 2013, and any Other Services to be provided.

Copyright Licence:
- ✔ This document is for the sole use of the purchaser.
- ✔ You may edit and amend it to match the style and procedures of your practice and as required for specific projects.
- ✔ You and the other parties to your contracts may distribute copies of this document to advisers and to other persons as necessary in connection with the proper performance of your contracts.
- ✘ In any other case, you may not distribute or reproduce the whole or any part of this document in any work, whether in hard copy, electronic or any other form, without the prior written consent of RIBA Enterprises Ltd.
- ✘ You may not use the RIBA initials or logo unless you are an RIBA chartered member or RIBA chartered practice.

Further guidance
A *Guide to RIBA Agreements 2010 (2012 revision)* explains fully how the RIBA Agreements 2010, 2012 revision work and their legal context.

Copyright notice

© Royal Institute of British Architects, 2013
Small Project Services Schedule, 2013 (RPoW Compatible)

Published by RIBA Publishing, 15 Bonhill Street, London EC2P 2EA
First published 2010; updated September 2011
Published 2013

RIBA Publishing is part of RIBA Enterprises Ltd.
www.ribaenterprises.com

Small Project Services Schedule, 2013

☐	Identify extent of design work by contractor
☐	Prepare documents required for tendering purposes
☐	Advise on potential contractors to be invited to tender for the work
☐	Invite, appraise and report on tenders and prepare the Building Contract and arrange for it to be signed
☐	Provide the contractor with information required for construction
☐	Review design work produced by contractor's specialist subcontractors
☐	

CONSTRUCTION

Stage 5 – Construction

☐	Visit the site to see that the work is proceeding generally in accordance with Building Contract
☐	Certify payments for work carried out or completed. Advise on final cost
☐	Provide or obtain 'As Constructed' information showing the building and its services
☐	Give general advice on maintenance
☐	

Stage 6 – Handover & Close Out

☐	Make final inspections and advise on resolution of defects
☐	Agree final account and issue a final certificate
☐	

AFTER CONSTRUCTION

Stage 7 – In Use

☐
☐

Other Services

☐
☐
☐

The following activities do not form part of the Services unless identified as 'Other Services' above:

- ○ Models and special drawings
- ○ Negotiating approvals by statutory authorities
- ○ Making submissions to and negotiating approvals by landlords, freeholders, etc
- ○ Preparing a schedule of dilapidations
- ○ Services in connection with party wall negotiations
- ○ Negotiating a price with a builder (in lieu of tendering)
- ○ Dealing with extensions of time and contractor's claims
- ○ Services in any dispute between the Client and another party
- ○ Services following damage by fire and other causes
- ○ Services following suspension, termination of any contract or agreement with or the insolvency of any other party providing services to the project
- ○ Services in connection with government and other grants

Small Project Services Schedule, 2013

Delete or strike through any services not required. Check the boxes for the required services [Y or ✔], alternatively enter 'T' for time-based services or 'LS' for lump sums, make any necessary amendments and/or add additional services.

The Services shall be performed in accordance with this Schedule in stages: [ ]

BEFORE CONSTRUCTION

Stage 0 – Strategic Definition
- [ ] Assist the client in preparing the Strategic Brief

Stage 1 – Preparation & Brief
- [ ] Visit the property and carry out an initial appraisal
- [ ] Assist the client in developing the Initial Project Brief
- [ ] Prepare and discuss alternative Feasibility Studies for the project
- [ ] Establish the clients Project Budget
- [ ] Advise on the project roles/other members of the project team required to carry out the project

Delete (a) or (b)
- [ ] (a) Make or (b) Arrange: survey of site and/or buildings
- [ ] Arrange investigation of soil or structural conditions
- [ ]

Stage 2 – Concept Design
- [ ] Prepare a preliminary Concept Design and discuss with the client
- [ ] Develop the Concept Design and Final Project Brief
- [ ] Provide updated Cost Information
- [ ] Submit the Concept Design, Final Project Brief and Cost Information for approval
- [ ]

Stage 3 – Developed Design
- [ ] Develop design with rest of project team
- [ ] Provide updated Cost Information
- [ ] Submit the Developed Design and Cost Information for approval
- [ ] Make an application for detailed Planning Permission
- [ ]

Stage 4 – Technical Design
- [ ] Develop the Technical Design in sufficient detail to enable a tender or tenders to be obtained
- [ ] Co-ordinate and integrate the designs of the other project team members
- [ ] Make an application for Building Regulations approval

Delete (a) or (b)
- [ ] Prepare (a) a specification or (b) a schedule of works
- [ ] Advise on an appropriate form of Building Contract, its conditions and the responsibilities of the client, the design team and the contractor

RIBA

RIBA Plan of Work 2013

Small Project Services

Schedule 2013
(RIBA Plan of Work 2013 compatible version)

This is the Small Project Services Schedule 2013 referred to in the Agreement relating to

The Project, namely:

between

The Client, namely:

Initials

and

*

Initials

---

* Insert 'The Architect' or 'The Consultant', namely

Identification Box

The identification box above is not always necessary. For example, if you are incorporating this schedule into a document which already has an identification box, there is no need to include it.

However, if you decide to use it, you should amend the text so that it is clear what project it relates to.

Note that if you are adding this document as an Appendix to your Agreement, you should use the identification box to say so. For example, you could replace the top line of text with 'This is Appendix <insert reference> referred to in the Agreement relating to:'.

Alternatively, where the law of Scotland applies, amend the top line to read 'This and the following <number> pages (numbered 2 to <number>) is the <accurate title of document> referred to in the Agreement relating to:'

If used with the Sub-consultant Agreement, 2012 revision, amend 'The Client' to 'The Consultant' and 'The Architect' or 'The Consultant' to 'The Sub-consultant'.

Royal Institute of British Architects

Concise Agreement 2010 (2012 revision): Notes: Part 2 Model Letter, 2012 revision

Refer to note 10
AGREEMENT

This Agreement is subject to the law of [England and Wales] [Northern Ireland] [Scotland].

The Client, <insert name>, wishes to appoint the [Architect] [Consultant] <insert name> for the Project and [Architect] [Consultant] has agreed to accept the appointment.

It is agreed that in accordance with the terms of this Agreement, the [Architect] [Consultant] performs the Services, and the Client pays [Architect] [Consultant] for the Services and performs the Client's obligations.

This Agreement comprises this Letter of Appointment and the attachments listed below, each identifying the Project, the Client and the [Architect] [Consultant] and initialled by the parties before signing this Agreement.

The *Concise Conditions of Appointment for [an Architect] [a Consultant], 2012 revision* with schedule *[Small Project Services]* [<insert other schedule>]

Appendix <insert reference> 'The Brief'

Appendix <insert reference> Notes of discussions on the Conditions of Appointment

Appendix <insert reference> Amendments to the Conditions

Appendix <insert reference> Fees and Expenses Schedule *Available as an electronic component, part of RIBA Agreements 2010 (2012 revision).*

Appendix <insert reference> Supplementary Conditions for a Public Authority

<insert any other appendices required>

This Agreement was made as a simple contract[2] on _____ of _____ 20__

Signed _____   _____

          Client   [Architect] [Consultant]

_____   _____

Witness signatures if required by the parties[3]

---

[2] If this Agreement is to be signed as a deed, delete the above and insert appropriate provisions to suit the status of the parties and their signatories. For examples see *Guide to RIBA Agreements 2010 (2012 revision)* or the electronic version of these Notes.

[3] If required by the parties where the law of England and Wales or the law of Northern Ireland applies, or to be self-proving under Scottish law, (subject to compliance with S.3 of *Requirements of Writing (Scotland) Act 1995*).

Until the expiry of the liability period, professional indemnity insurance cover will be maintained for that amount to be available for your Project except for claims arising out of:
- pollution and contamination, where the annual aggregate limit is <insert amount>.
- asbestos and fungal mould, where the limit for any one claim and in the aggregate is <insert amount>.

Documentary evidence of the insurance can be provided, if required.

Refer to note 9
Disputes

My practice aims to provide a professional standard of service, but if at any time you are not satisfied, please bring the issue to my attention as soon as possible and we can discuss how to resolve the issue. We hope we shall be able to settle the matter by negotiation or mediation.

However, either of us can refer the matter to adjudication under [procedures published by the Construction Industry Council] [the relevant statutory Scheme for Construction Contracts Regulations] [the RIBA Consumer Contracts Adjudication scheme]. Should we need help in choosing an adjudicator, the nominator will be <name of selected body>.

We have also agreed that, without prejudice to that right of adjudication, a dispute or difference may be referred [to legal proceedings and clause 9.3 will be deleted] [to arbitration in accordance with the provisions of clause 9.3. Should we need help in choosing an arbitrator, the appointor will be the] [President or a Vice President of <name of selected body>] [President or the Secretary of the Royal Incorporation of Architects in Scotland.]

Completing the Agreement

I confirm that performance of the Services [will commence] [commenced] on <insert date>.

I also confirm that my practice will be represented by <insert name> and that you will be represented by <insert name>, who will have sole authority to act on your behalf for all purposes under the Agreement.

Refer to note 10
If these arrangements are acceptable to you, please sign the Agreement clause [below] [overleaf], initial each of the appendices where indicated, and return all the documents. They will then be countersigned and the date entered in the Agreement clause, and a certified copy set will be sent to you for your records.

Refer to note 11
I am enclosing, for your information, a copy of *A Client's Guide to Engaging an Architect 2009*, which includes a brief outline of some relevant legislation and *A Client's Guide to Health and Safety for a Construction Project*, which outlines the obligations of a business client under CDM, *(Not applicable to domestic projects.)* which I hope you will find helpful.

Refer to note 12
[I am also enclosing] [I confirm that I have provided you with] background information about my practice [is available] [on request] [<at website address.>]

Yours faithfully

for and on behalf of <practice name>

Concise Agreement 2010 (2012 revision): Notes: Part 2 Model Letter, 2012 revision

Refer to note 5
Services

[The Services to be provided are <describe the professional services required>.]

[The Services will be provided in accordance with the Small Project Services Schedule].

[The Services are set out in the schedule of Services – Role Specifications, Design and Other Services.] In addition to our role as Designer, the Services include performance of the roles of Lead Consultant Contract Administrator Lead Designer CDM Co-ordinator. *(Not applicable to domestic projects.)*

If it becomes necessary to vary the services, we can discuss how this might be arranged.

Refer to note 6
[At this time I do not believe it will be necessary to seek advice from any other consultants, but if this should change I will advise you about the requirements and the fees entailed.]

[If it should prove necessary to seek advice from any other consultants or specialists, I will advise you at the appropriate time about such appointments and the fees entailed.] Please make the outstanding appointments as soon as possible. As the CDM Regulations apply to this Project, you must appoint the CDM Co-ordinator as soon as RIBA Stages A and B are complete. *(Not applicable to domestic projects.)*

[I informed you that the services of a [structural engineer] [quantity surveyor] will be required. I will write separately about their appointment and the fees entailed.]

[As agreed with you, <insert name>, a firm of [structural engineers] [quantity surveyors] will be engaged by my practice, who will be responsible to you for their services, the costs of which are included in my/our fee.]

Refer to note 7
Fees

Our Fees are set out [below <insert description> *(See under Fees and Expenses in Guide to RIBA Agreements 2010 (2012 revision).)*] [in the *Fees and Expenses Schedule, 2012 revision*, Appendix <insert reference>.]

Before implementing any changes required to the Services or an approved design, the basis for any consequential change to the fees or expenses will be agreed with you.

Accounts will be submitted [monthly] [on completion of each stage] for fees and other amounts due. Please note clauses 5.8 and 5.9 are derived from statutory provisions and *(Applies to business clients only.)* that the final date for payment of our account is 28 days after the date of issue. Any sums remaining unpaid after 28 days will bear interest plus reasonable debt recovery costs as set out in clause 5.12.

[In addition to fees, the expenses listed below will be charged [at net cost] [plus a handling charge of <insert percentage> of net cost] [by adding <insert percentage> of the total fee to the amount due.] [<list expenses>]
[The fee includes expenses.]

Any disbursements made on your behalf, such as payments to the local authority for planning submissions or Building Regulations approval, will be charged at [net cost] [plus a handling charge of <insert percentage>]

[My practice is registered for VAT, which is chargeable on all fees and expenses.]

[My practice is not registered for VAT but if during the course of this appointment it is necessary to register for VAT, this will change.]

Refer to note 8
Liability

We discussed the potential risks associated with your project and agreed that liability to you for loss or damage, subject to clauses 7.2.1 and 7.3 *(See under Fees and Expenses in Guide to RIBA Agreements 2010 (2012 revision).)* will be limited to <insert amount> [in the aggregate (the overall cap for all claims)] [in respect of each and every claim or series of claims arising out of the same originating cause.]

Concise Agreement 2010 (2012 revision): Notes: Part 2 Model Letter, 2012 revision

Refer to the numbered notes and adapt the model text as appropriate. The symbols < > indicate where project specific information should be inserted. The sections of text highlighted in grey are optional and can be deleted or varied as appropriate. The sections of text in [square brackets] are alternatives, generally only one of the sections should be selected.

Refer to note 1
<insert your full business name and address>

<insert Client's full business name and address>

For the attention of <insert name>

Refer to note 2
Dear [Sir] [Madam]

Project: <project title>

at <site address> ('the Site')

Thank you for inviting my practice to act as your [Architect] [Consultant] for this project.

I am now writing to confirm our discussions.

Refer to note 3
[Your letter dated <insert date> attached as Appendix <insert reference> sets out your requirements and information about the Site – 'the Brief']. [Following our meeting and our visit to the site I have set out your requirements and information about the Site – 'the Brief' below in Appendix <insert reference>]

You told me that your target cost for the building work is <insert amount>, to which must be added our fees and those of any other consultants, together with any VAT. You also said that you would like building works to [commence on] [be complete by] <insert date>.

We have agreed that the Agreement with you will comprise this Letter of Appointment together with the *RIBA Concise Conditions of Appointment for [an Architect] [a Consultant], 2010 (2012 revision)* incorporating the proposed amendments to the Conditions, Appendix <insert reference> and other documents referred to in this letter.

Refer to note 4
This paragraph applies to consumer clients only
I also confirm that I drew your attention to and explained the purpose of each of the clauses. This letter records the results of our discussions [as shown in Appendix <*insert reference*>] and in particular our decisions on:
- clause 5 relating to fees, expenses and payment. You agreed to follow the procedures set out subject to deletion of clause 5.9 *(Clause 5.9 is derived from statutory requirements which are not obligatory for 'residential occupiers')* and the reference to it in clause 8.1 and deletion of clause 5.10;
- clause 7 relating to liability. You asked for clause 7.3 (net contribution) to be [deleted] [amended to provide an overall cap on liability] and agreed that the time limit and the amount of professional indemnity cover specified below appeared to be reasonable for your project;
- clause 9.1 relating to the resolution of any dispute.

You also noted the provisions of clause 10.

When the contract documents are ready, send the Conditions of Appointment, the schedule, any appendices and the Letter of Appointment to the Client. It might be helpful to flag places where signatures or initials are required.

Note that in the identification box on the various components, the name of the parties must identify their legal identities.

If a consumer Client has acknowledged clause 10, there is no requirement to take further action.

Either:

Send the documents to the client asking for all the documents to be returned, when signed and initialled.

On return of the documents, check that all is in order, and sign and initial the documents. Enter the date of the Agreement and send a certified true copy to the Client for record purposes. The Architect/Consultant retains the originals.

A certified true copy may be a photocopy of the signed and initialled documents, each of which indicates, perhaps using a pre-printed adhesive label on the face:

> 'This is a true copy of the document referred to in the Agreement
>
> between <Client> and <Architect/Consultant> dated <date>
>
> signed <signature of Architect/Consultant as original>

Or

Send the Client's copy of the documents to the Client for consideration and arrange a meeting when Client and Architect/Consultant can meet to sign and initial the documents. Don't forget to take the Architect/Consultant's copy to the meeting.

See also *Guide to RIBA Agreements 2010 (2012 revision)* on completion and attestation in England, Wales and Northern Ireland and the special requirements in Scotland where the letter of appointment must comply with the *Requirements of Writing (Scotland) Act 1995* to be valid.

11 Provision of a copy of *A Client's Guide to Engaging an Architect*, which includes a brief outline of the legislation and fee options and *A Client's Guide to Health and Safety for a Construction Project*, which outlines the obligations of a business client under CDM, may help the less experienced Client to understand what an Architect can be expected to do and charge. Both titles are published by RIBA Publishing.

12 This may also be an appropriate place to draw attention to the availability of the information about the practice required by the *Provision of Services Regulations 2009*, if not already provided to the Client.

8 It may be beneficial to discuss the professional indemnity insurance cover for the project with the insurers/broker, particularly if the net contribution condition is deleted or the Client is a consumer or sub-consultants are proposed.

   Under clause 7.2 the Architect's/Consultant's liability for loss or damage does not exceed the amount of the Architect's/Consultant's professional indemnity insurance. The amount of that insurance available for the project, which is recorded in the Letter, should be reasonable in relation to the risks to pass the reasonableness test under the *Unfair Contract Terms Act 1977*, but it can be less than the cover carried by the Architect's/Consultant's practice.

   The related net contribution clause 7.3 is designed so that the Architect/Consultant pays a fair share of the loss or damage to the claimant on the assumption that any other parties to the loss have also paid their share. If the Client requires the clause to be deleted, it might, after discussion with the insurers/broker, be replaced with 'The total liability of the [Architect] [Consultant] under or in connection with this Agreement shall not exceed <amount>.' This may affect the professional indemnity insurance premiums payable and in turn the amount of the fee required.

   However, in the case of a consumer Client, such a condition effectively eliminates the legal right of 'joint and several liability' and replacing it with a cap on liability may offend against the *Unfair Terms in Consumer Contracts Regulations*. Nevertheless, in the case of *James Moores v Yakeley Associates Ltd*, the court accepted that the cap on liability was reasonable.

   CIC Liability Briefings on *Net contribution clauses* and *Managing liability through financial caps* are available at *www.cic.org.uk/liability*, but following the Scottish case of *Langstane v Riversides* [2009] CSOH 52, it may be that this issue will need to be revisited.

   Architects must maintain insurance for not less than the amount required by the Architects Registration Board and include cover for legal defence costs. PI insurance cover should run for at least the period stated in clause 7.1.

9 *Guide to RIBA Agreements 2010 (2012 revision)* covers selection of the dispute resolution procedures.

   In a contract with a business client both parties have a statutory right to adjudication. Select the procedures published by the Construction Industry Council (*www.cic.org.uk*) or the relevant statutory Scheme for Construction Contracts Regulations (*www.legislation.gov.uk/uksi*). Choose between legal proceedings and arbitration. Note also that the joinder provision under CIMAR rule 2.6 will apply only where the contracts of other parties to the dispute also include an arbitration condition. A consumer has the right to refer any dispute to the courts. Any other options must be negotiated. For instance, if the consumer wishes to include arbitration provisions, perhaps to keep the matter private, the clauses would need to be amended by including a clause such as 9.3 in the Standard Conditions. Note that requiring a consumer to take disputes exclusively to arbitration is likely to be considered unfair (UTCCR Schedule 2.1(q)).

   Alternatively, if a consumer chooses adjudication, the RIBA Adjudication Scheme for Consumer Contracts, which may be suitable for low value claims, could be selected perhaps in addition to or as an alternative to the procedures published by the Construction Industry Council or the relevant *Scheme for Construction Contracts Regulations*. Details of the RIBA Scheme are available from the Disputes Resolution Office, T: 020 7307 3649 F: 020 7307 3793 adjudication@inst.riba.org

   Only if the negotiations to agree the dispute procedures are recorded (see note 4) can an Architect/a Consultant be assured that the decisions of an adjudicator/arbitrator could not be challenged by a consumer client on the grounds of no jurisdiction.

10 If the Agreement is being completed after performance of the Services has commenced, ensure the entries in the various components are compatible with the basis on which the Architect/Consultant started work.

   Note that if the Agreement is not signed by a consumer client before work has started, the consumer's right to cancel in the seven day period after signing it may give the consumer time for second thoughts after the Architect/Consultant has invested time and creative energy.

Concise Agreement 2010 (2012 revision): Notes: Part 2 Model Letter, 2012 revision

The Letter of Appointment is to confirm what has been agreed or is proposed. When drafting the Letter, refer to the numbered notes, the *Guide to RIBA Agreements 2010 (2012 revision)* and adapt the text as appropriate, particularly where the Client is a consumer.

The style of the Letter is less important than its contents, which are necessary to complete the terms of the Agreement. However, if the model text is not used, be careful that other wording does not modify or conflict with other provisions of the Agreement. The essential requirement is that all the necessary information is recorded in the Letter and/or in an appendix. Omission of any of the model contents should be considered carefully.

Numbered notes

1 For purposes of identification, insert accurate names and addresses of the parties to the agreement.

2 The Letter should be formal – 'Dear Sir …Yours faithfully … for and on behalf of [the practice]'.

 Ensure the title and location of the project are accurately stated.

 For a consumer Client it might be – 'Dear Mr and Mrs' and 'Yours sincerely … for and on behalf of [the practice]', but avoid social or other comment that might subsequently cause conflict with what is intended to be a legal agreement.

3 The brief should include the Client's initial statement of requirements on which the Services and Fees will be based, and include information about the Site. Under CDM 2007, a business Client 'whether or not the Project is notifiable' is required to provide 'all the information in the Client's possession, or which is reasonably obtainable, and which is necessary for the proper and timely performance of the Services including any such information in a health and safety file, or other information about or affecting the site or the construction work'.

4 Insert a paragraph to record the substance of the negotiations and any amendments that are agreed, perhaps in an appendix.

5 The Services can be specified in the Small Project Services Schedule or, using the editable electronic version, in a customised schedule, eg if the original is not suitable because the commission does not relate to a typical building project or a different plan of work is used. Alternatively, one of the other electronic schedules may be suitable.

6 So that the Client is clear about the extent of the appointment and understands his/her role in relation to the appointment of other consultants, the Letter should also refer to any other appointments which may be necessary. Whilst the option for employing sub-consultants is referred to, the Architect/Consultant should carefully consider the risks of doing so.

7 Fees and expenses can be specified in the Letter, as indicated in *Guide to RIBA Agreements 2010 (2012 revision)*, or in a fees and expenses schedule, for which an electronic version is available.

 Indication of the fee for each stage or group of stages will prove beneficial if changes are made subsequently to the cost or programme.

 Unless the fee for the construction stage is time-based, it is advisable to state the number of visits to the site in the construction stage included in the quoted fee as a basis for negotiations if a greater number of visits proves to be necessary.

 As appropriate, draw the Client's attention to any items in the Services for which time charges or a lump sum will apply and check that there is no inconsistency between the wording of the Letter and the specified services.

 Carefully list the expenses to be reimbursed – thus by implication defining those not covered. Where the net cost option applies, state the rates for copies made in the office and the mileage rate for travel by car.

Negotiating the terms of an Agreement with a consumer Client

If the Client is a 'consumer' as defined under UTCCR, it is recommended that you read through the terms of the Agreement with the Client and individually negotiate each term in the context of the consumer's rights. Failure to do so could lead to certain terms of the Agreement being invalidated.

UTCC Regulation 5.1 states: 'A contractual term which has not been individually negotiated shall be regarded as unfair if, contrary to the requirement of good faith, it causes a significant imbalance in the parties' rights and obligations arising under the contract, to the detriment of the consumer.'

The negotiations should enable the consumer to 'influence the substance' of the terms and minimise the risk that subsequently they would be considered to be unfair to the Client. It is not unheard of for a consumer to claim subsequently that they did not understand the implications of a term.

Schedule 2 to UTCC Regulation 5.5 gives an 'indicative and non-exhaustive list of terms which **may be** regarded as unfair.' The list includes terms having the effect of 'inappropriately excluding or limiting the legal rights of the consumer' or 'excluding or hindering the consumer's right to take legal action or exercise any other legal remedy'.

Whilst RIBA Conditions of Appointment are designed as an entity and provide remedies for the client in the event of default by the Architect/Consultant, eg the requirement to maintain PI insurance and the options for dispute resolution, some of the terms may need careful explanation in the negotiations:
- statutory requirements not applicable to 'residential occupiers' or 'consumers' (clauses 5.8, 5.9, 5.11 and 5.12) although these are mainly procedural;
- exclusion of the right of set-off (clause 5.10);
- recovery of costs (clause 5.13.1);
- the contractual limitation period and the limitation of liability in amount (clauses 7.1 and 7.2);
- exclusion of the right of joint and several liability (as in the net contribution clause 7.3) – see also numbered note 8;
- dispute resolution (clause 9) – see also numbered note 9.

If the Client requires amendment to or deletion of any of these provisions, the Architect/Consultant will need to consider any change to liability.

The following numbered notes provide some indication of the issues to be addressed. Note 4 draws attention to the need to record the substance of the negotiations and any amendments that are agreed, so that it can be shown that the Agreement accords with the Regulations.

If the negotiations are protracted or sending the Letter is delayed, it may be appropriate to send a note of the discussions to the Client immediately after the meeting(s).

If the Client is a married couple or joint residential occupiers, all the Client parties are consumers, but the Client should identify one of their number as their representative with full authority to act on behalf of the parties and to sign the agreement.

For work to the Client's home or to a second home, the client will be exempt from any statutory duties arising from:
- *the Housing Grants, Construction and Regeneration Act 1996* (HGCR) as amended by the *Local Democracy Economic Development and Construction Act 2009* as a 'residential occupier';
- *the Construction (Design and Management) Regulations 2007;*
- *the Late Payment of Commercial Debts (Interest) Regulations 2002.*

However, if the Client's second home is to be let at any time as a holiday rental or to other tenants, the Client will be deemed to be a business client and the exemptions will not apply.

Clause 10 extends that right to cancel an agreement, to wherever it was signed and to any agreement, including one relating to the construction of a new home.

Guidance on 'Negotiating the terms of an Agreement with a Consumer Client' is given with the Model Letter.

A company may also be a 'consumer', subject to *Unfair Contract Terms Act 1977* but not the *Unfair Terms in Consumer Contracts Regulations 1999* (UTCCR) if the transaction is 'only incidental to its business activity and which is not of a kind it makes with any degree of regularity.' It may be a wise precaution and courteous to treat such consumers as though the UTCCR did apply.

A 'consumer' who is not a 'residential occupier' will **not** be exempt from HCCRA as amended or exempt from the CDM Regulations if not a 'domestic client' as defined in the CDM Regulations.

The Schedules of Services

The included schedule *Small Project Services* (also available online) covers design and management roles.

Tick the boxes in the printed schedule for the Services required or enter 'T' for time charged services or 'LS' for lump sums. The schedule relates to a straightforward project to be procured in the traditional manner. Although the Services are described in simple terms, performance must be 'in accordance with the normal standards of the Architect/Consultant's profession' (Clause 1). If different or extra services are needed for the particular project, these can be described and identified as 'Other Services'.

The Services should accurately reflect the Client's requirements, particularly where the Architect/Consultant is to perform other professional services; or only part of the Services within a work stage, or only preparing and submitting a planning application.

If the schedule is not suitable because, for instance, the commission does not relate to a typical building project or a different plan of work is used, it can be removed easily from the centre fold in the paper version, it can be replaced at the back of the Conditions with a customised project specific version, perhaps using the editable on-line version of the schedule or a project specific schedule should be devised.

If it is obvious that further details are required, consider how these are to be obtained.

Online schedules of services available include:
Access Consultancy Services
Historic Building or Conservation Project Services
Initial Occupation and Post-occupation Evaluation Services
Master Planning Services
Small Historic Building or Conservation Project Services
Small Project Services

For some commissions it might be sufficient to include the details of the services required in the Letter of Appointment, eg for a specialist activity such as an accessibility audit, for an advisory service, or investigation of a building failure.

For more complex projects, particularly where the Project is notifiable under the CDM Regulations and/or other consultants are engaged, the schedule *Role Specifications, Design and Other Services* might be more appropriate. The Role Specifications outline the responsibilities and authority of the Lead Architect/Consultant, Lead Designer, Contract Administrator etc and defines the boundaries between different appointments.

Fees and Expenses

This optional component provides a structured basis for recording the basis of the fee and alternative arrangements for reimbursement of expenses.

Concise Agreement 2010 (2012 revision): Notes: Part 1 Use and Completion, 2012 revision

> **A Concise Agreement** is a suitable basis:
>
> - for a commission where the concise contract terms are compatible with the complexity of the project and the risks to each party; and
>
> - where the Client is acting for business or commercial purposes; or
>
> - where the commission is for work to the Client's home and the terms have been individually negotiated with the Client as a 'consumer'.
>
> - the building works, including extensions and alterations, will be carried out using forms of building contract[1], such as JCT/SBCC Minor Works Building Contract, or JCT Intermediate Form of Building Contract.

RIBA Concise Agreement 2010, 2012 revision will comprise the *Concise Conditions of Appointment, 2012 revision*, a schedule of services, any appendices and a Letter of Appointment, which is signed under hand as a simple contract or is, where the law of Scotland applies, validly subscribed.

The Architect version of the Conditions, the Schedules and these Notes are available in print and as electronic files. The Consultant version of the Conditions is only available electronically. Electronic versions are available at *www.ribabookshops.com/agreements*

The Provision of Services Regulations 2009 defines the minimum amount of information which service providers must make available to clients including the registered status of service providers (architects or consultants) who will be party to the Agreement. Some of the information is included in RIBA Conditions of Appointment. The Model Letter of Appointment also includes suggested wording before the signature about the availability of the information, which can be supplied on your own initiative, be easily accessible at your business address, or held on a publicly available weblink, if not already provided to the Client.

The Conditions

The obligations under this Agreement are similar to those under the RIBA Standard Conditions of Appointment, and include the relevant statutory obligations. However, some of the rules or procedural requirements in the Standard Conditions do not appear. It is, of course, implicit that 'normal standards' are consistent with the requirements of the Architect/Consultant's professional code of practice. For complex or high value projects, a RIBA Standard Agreement may be more appropriate.

The Conditions of Appointment set out in concise terms the obligations of the Architect/Consultant and may be used where the Client is acting for business or commercial purposes or is a Public Authority.

An Agreement with a business client or a public authority is a 'construction contract' to which the *Housing Grants, Construction and Regeneration Act 1996* (HGCRA) as amended by Part 8 of the *Local Democracy, Economic Development and Construction Act 2009*, applies.

Business clients include charities, religious organisations and not-for-profit bodies. Where the Client is a public authority, it may be necessary to include provisions relating to the *Freedom of Information Act 2000* and to corrupt gifts and payments, for which an electronic Public Authority Supplement is available as part of RIBA Agreements 2010, 2012 revision.

The Conditions may also be used for work to the Client's home or a second home, including a new house or apartment, if the Client has elected to complete an Agreement in his/her own name, ie not as a limited company or other legal entity. The Client will be:
- a 'consumer' ie 'a natural person acting for purposes outside his trade, business or profession' to whom the *Unfair Terms in Consumer Contracts Regulations 1999 (SI 2083) (UTCCR);* and
- *The Cancellation of Contracts made in a Consumer's Home or Place of Work etc Regulations 2008* (SI 1816) apply.

The Cancellation of Contracts made in a Consumer's Home or Place of Work etc Regulations 2008 establishes the right of a consumer to cancel an agreement for any reason within seven days from the date when the Client signed it.

---

[1] Other contracts include *ICE Conditions of Contract for Minor Works or CIOB Small Works Contract.*.

The Notes are in two parts, Part 1 covers Use and Completion of the Agreement and Part 2 provides a Model Letter of Appointment to assist the drafting of a project specific letter. The Model Letter concludes with an Agreement clause to be signed by both parties as a simple contract. Alternatively guidance is given if the Model Letter is to be signed as deed.

In this updated edition amendments have been made on page 3, including:
- to the note on the *Provision of Services Regulations 2009*
- noting that the *Housing Grants, Construction and Regeneration Act* is amended by Part 8 of the *Local Democracy, Economic Development and Construction Act 2009*.

These notes may be read in conjunction with *Guide to RIBA Agreements 2010 (2012 revision)*, which gives general guidance on getting started; terms of RIBA Agreements; consumer clients; the brief; the services; watch points – fees and expenses, copyright, liability and insurance, net contribution, dispute resolution, suspension and termination; final details, amendments and attestation.

All parties must rely exclusively upon their own skill and judgement or upon their advisers when using electronic components from the suite of RIBA Agreements 2010, 2012 revision and neither RIBA nor RIBA Enterprises Ltd assume any liability to any user or any third party.

Copyright Licence
- ✔ This document is for the sole use of the purchaser.
- ✔ You may use it for one professional services contract only.
- ✔ You and the other party to your contract may distribute copies of this document to advisers and to other persons as necessary in connection with the proper performance of your contracts.
- ✘ In any other case, you may not distribute or reproduce the whole or any part of this document in any work, whether in hard copy, electronic or any other form, without the prior written consent of RIBA Enterprises Ltd.
- ✘ You may not use the RIBA initials or logo unless you are an RIBA chartered member or RIBA chartered practice.

Copyright notice

© Royal Institute of British Architects, 2010, 2012
Concise Agreement 2010 (2012 revision): Notes: Part 1 Use and Completion; Part 2 Model Letter

Published by RIBA Publishing, 15 Bonhill Street, London EC2P 2EA
First published 2010; updated September 2011
Revised edition published 2012

RIBA Publishing is part of RIBA Enterprises Ltd.
www.ribaenterprises.com

RIBA

Concise Agreement 2010: Notes:

Part 1 Use and Completion;
Part 2 Model Letter

(2012 revision)

Royal Institute of British Architects